J. R. S. Sterrett

Preliminary Report of an Archaeological Journey

Made in Asia Minor During the Summer of 1884

J. R. S. Sterrett

Preliminary Report of an Archaeological journey
Made in Asia Minor During the Summer of 1884

ISBN/EAN: 9783744746380

Printed in Europe, USA, Canada, Australia, Japan

Cover: Foto ©berggeist007 / pixelio.de

More available books at **www.hansebooks.com**

PRELIMINARY REPORT

OF AN

ARCHÆOLOGICAL JOURNEY

MADE THROUGH

ASIA MINOR

DURING THE SUMMER OF 1884,

BY

J. R. S. STERRETT.

INTRODUCTORY NOTE.

THIS Preliminary Report of Dr. Sterrett's journey in Asia Minor in 1884 will form a part of the Second Volume of PAPERS OF THE AMERICAN SCHOOL OF CLASSICAL STUDIES AT ATHENS, which will be published, it is hoped, in the summer or autumn of 1885. It is now issued separately, in the hope that the collection of inscriptions, among which are those of forty-two Roman milestones found in Cappadocia, will be of value to all who are interested in the geography of Asia Minor. Dr. Sterrett announces his intention of presenting these inscriptions hereafter in cursive text, with historical and explanatory notes (see page 20).

Dr. Sterrett was a student of the American School at Athens during its first year, 1882–1883, and he kindly returned to it the next year to assist Professor Packard. This journey in Asia Minor was aided by the contributions of several gentlemen in Boston and Cambridge.

This publication precedes by a few weeks the First Volume of Papers of the School.

WILLIAM W. GOODWIN, } *Editors.*
THOMAS W. LUDLOW, }

January, 1885.

To the Managing Committee of the American School of Classical Studies at Athens.

Athens, Oct. 15. 1884.

During the few days that remain to me before my departure on the Wolfe expedition to the Tigris-Euphrates valley, it is possible to lay before you only a brief outline of the results of my journey through Asia Minor during the past summer. In this outline I shall call your attention to important facts alone, leaving untouched many things well worthy of notice.

Early last spring I laid before you an agreement between Mr. W. M. Ramsay of Exeter College, Oxford, and myself, concerning a joint journey to be made in Asia Minor. We agreed to work in concert through a given district for three or four weeks, after which we were to separate and carry on our summer's work independently. In pursuance of this agreement we met in Smyrna, May 15, 1884, where I provided myself with the outfit which would be necessary for my own journey after our final separation. I then went to Aïdin Giuzel Hissar, the ancient Tralleis, to buy horses and make other final arrangements. For the present I had need of only four horses : one for myself, two for my two personal attendants, and one baggage horse, which, besides its burden, had to carry the cook. These were easily procured.

Mr. Ramsay being necessarily delayed in Smyrna, I undertook an excursion in the direction of Nazli, and found near Kiösk a fragmentary letter of one of the later kings, insignificant in itself, but of value in so far as it locates approximately Ἱερὰ Κώμη, which has heretofore been placed on the west of Tralleis.

No. 1.

```
\U_.._IU\/\I_
OΣTOҰENTHIEPAKΩιιιι        TEKAIΩΣETIMHΘHΔIATAΥ
ΖΩKENAΞIΩMAΔIOҰEΛE         THNΠATPIONBA▨▨ΛEIANKAΙ
IIΣONIEPAΣKOMHΣKATOIʺʺ      TOΣTETAYΠOT
IΔPҰM ENATΩ AΠ OΛΛΩN I      ΣKHΠTPONEXOYΞHIKᴗ
ΣTAΣTOҰΘEOҰΘEPAΠEIΛΞ        ҰNT=ΛEINKAΘҰΔ
ΖΣAΠAPXHΣEIXENEΓΩΔE
ΠOTΩNΠPOEMOYBAΣI
VΞ EINTEKAITATΩNΘE
        THN
```

Our final start was made from Kuyndjak by way of Antiochia to Aphrodisias. Antiochia has disappeared entirely, and from the villages of this region we collected only a few insignificant inscriptions. The ruins of Aphrodisias are, on the contrary, very extensive. A vast number of inscriptions from this place are already known; and, in our best judgment, we should have required at least a fortnight to work our way through the wilderness of ruins and sift the new from the old, the known from the unknown. Accordingly, as time was pressing, we reluctantly postponed a minute investigation of Aphrodisias until a more convenient season. But, even after this hurried visit, I venture to express my belief that systematic excavations on this site would yield ample returns for the outlay.

From Aphrodisias Mr. Ramsay went around Baba Dagh to the north, by way of Denizli, and I to the south.

On Kiepert's large map of Asia Minor (1852–55) it will be found that the old site at Makuf is claimed for Trapezopolis; but at Makuf, besides numerous other inscriptions, I found one which shows that in future maps Heraclea must be inserted here instead of Trapezopolis.

No. 2.

```
HΘHKHHΓOPACΘHYΠOTITOYCTATIΛιᴗ
MHTIOXOYЄNHTЄΘHCЄTЄAYTOCKAIHΓYNι
AYTOYAYPHΛIAMЄΛITINHΔIONYCIOYK
ONANAYTOIΠЄPIONTЄCBOYΛHΘωCINЄTC
PωΔЄOYΔЄNIЄΞЄCTAIЄNΘAYЄTINAЄITι
```

```
)ΕΝΘΑΥΑΙΤΙΝΑΠΟΤΙСΕΙΤωΚΥΡΙΑΚω
ΦΙСΚωϪΦΚΑΙΤΗΒΟΥΛΗΤΗΗΡΑΚΛΕω░░
ΤωΝϪΦΚΕΟΥΔΕΝΗΤΤΟΝΟΕΝΤΑΟΟ
░░ΙΕΤΑΤ░░░ΘΗ░░░░░░░░░░░░░
░░░░░░░░░░░ΟΥΝΑΝΤΙΓΡΑΦΟΝΑΠΕ░░░
░░░░░░░░░ΕΙСΤΑΑΡΧΕΙΑ░░░░░░░░░
░░░░░░░░░░░░ΟСΕ░░░░░░░░░░░░░░
```

The Stadion at Heraclea is still very distinct. The Acropolis is a low hill of great extent on the top. The walls of the Acropolis are easily followed around the whole circuit. In some places they are level with the ground, while in others they are still erect. The walls have once been destroyed and afterwards rebuilt, as is clear from the architectural fragments and even inscribed stones which are built into the present wall. But that the foundations at least are chiefly antique is shown by the fact that on the outside the wall is provided with finely executed stone shoots at the bottom to carry the water off. Still, at one place, where the wall is now used as a quarry by the villagers of Makuf, I discovered an honorary inscription in the very foundation. The walls were evidently rebuilt in time of great and pressing need, when the anxious citizens made use of anything in the shape of stone that came in their way.

From Heraclea I zigzagged to the south-east and south through the plain now known as Davas Ovassi, and made a good survey of the district. I visited Tabae, now Davas; but found no inscriptions, and but few traces of a past other than Turkish. Tabae is situated on a high hill in a gorge between two mountains, and is surrounded by cañons three or four hundred feet deep on all sides except one. On this one side it is approached by a bridge which crosses a chasm where it is least deep; after the bridge is crossed a narrow neck of land, just wide enough for a roomy road, leads by a tortuous and laborious ascent to the town on the hill. When seen from any point in the plain it seems to be situated on a neck or saddle of the mountains; and one is extremely surprised at the real topography of the place.

The plain of Tabae is one of extraordinary fertility; in antiquity it supported three cities, Heraclea, Tabae, and a third at Medet, the name of which is as yet unknown to me. But that it was a town of considerable wealth is clear from a very substantial antique substructure in huge hewn stones of blue limestone. Upon these foundations there now stands the Mosque, which has without doubt superseded a prouder structure in honor of a pagan god. The adjoining hill, which certainly served as the Acropolis, contains no traces of antiquity, except the many architectural fragments of great weight and size which are found in the cemetery. At Medet, besides the ruins just mentioned, I found honorary inscriptions, which unfortunately do not give the name of the town.

From Medet I crossed over the mountains to Kizildje, where I found an inscription which settles the site of Sebastopolis. The inscription is given as it stands; and all errors must be laid at the door of the stone-cutter.

No. 3.

```
ΑΥΤΟΚΡΑΤΟΡΙΝΕΡΒΑΤΡΑΙΑΝΩ
ΑΡΙΣΤΩΚΑΙΣΑΡΙΣΕΒΑΣΤΩΓΕΡΜΑΝΙ
ΚΩΔΑΚΙΚΩΠΑΡΘΙΚΩ
ΚΑΙΤΩΔΗΜΩΤΩΣΕΒΑΣΤΟΠΟ
ΛΕΙΤΩΝ · Π · ΣΤΑΤΙΟΣΕΡΜΑΣΑΓΟ
ΡΑΝΟΜΗΣΑΣΚΑΙΠΑΡΑΦΥΛΑΖΑΣ
ΚΑΙΤΕΙΜΗΘΕΙΣΕΤΙΤΕΥΠΕΡ
ΤΗΣΣΤΡΩΣΕΩΣΤΗΣΣ Ζ̄Ζ̄Ε
ΔΡΑΣΤΗΣΕΝΤΩΤΕΤΡΑΣΤΥ
ΛΩΤΟΥΓΥΜΝΑΣΙΟΥΤΕΙΜΑΙΣ
ΕΙΡΗΝΑΡΧΙΚΑΙΣΠΑΛΙΝΔΕΥ
ΠΕΡΑΝΑΣΤΑΣΕΩΣΤΗΣΝΕΙ
 ΙΗΣΕΚΤΩΝΙΔΙΩΝΤΕΙΜΗΘΕΙΣ
ΤΕΙΜΑΙΣΔΙΑΝΥΚΤΟΣΣΤΡΑΤΗ
ΓΙΚΑΙΣΚΑΙΑΠΟΔΟΧΕΥΣΓΞ
ΝΑΜΕΝΟΣΓΧΜΑΣΚΑΙΑΡΓΥ
ΡΟΤΑΜΙΑΣ✱ΔΚΑΘΩΣΚΑΙ
ΔΙΑΤΩΝΥΗΦΙΣΜΑΤΩΝ
          ΠΕΡΙΕΧΕΙ
```

The ruins of Sebastopolis are in full view from Kizildje, about half an hour to the east. I had sent my cook in advance to Kizildje; but on my arrival there I found that he had gone to Kizildje Beylik, near Makuf; so, to my intense disgust, I had to leave Sebastopolis unvisited and go in search of the erring cook. Accordingly I re-traced my steps, leaving Medet on my left. From Kizildje Beylik the river Harpasus was traced to its source, the watershed being found to be east of Tekeh.

According to appointment, I had a conference with Mr. Ramsay at Kizil Hissar. He then took the plain of Karayuk Bazar, while I passed by Yataghan to Kayahissar, and thence through a number of villages (not on the map) to Güneh; thence to Eriza, the site of which is between Dodru Agha and Yazir. Here some interesting inscriptions were found. Hence I proceeded via Gumaoshar to Tchamkieui in the plain of Cibyra, where I again met Mr. Ramsay. Although we were so near, we decided not to visit Cibyra, knowing that it had been explored in 1877 by Messrs. Duchesne and Collignon, then of the French School at Athens. From Tchamkieui I returned northward to Derekieui. Half an hour to the north of Derekieui in the plain are foundations, possibly of a temple. On the top of the mountain immediately east of Derekieui the villagers report a *kale* and inscriptions. In the valley south-east of Derekieui are the ruins of an ancient town of some size, but no inscriptions. Hence I crossed the rugged and in places almost impassable Eshler Dagh to Karamanli. In this neighborhood, which contains the flourishing villages of Karamanli, Hedja, Sazak, and Tefeny, I spent three or four days copying inscriptions, which will be the subject of a special paper.

At Tefeny, where I again met Mr. Ramsay, some new inscriptions were found, one of which is very enigmatical, but is too long to be given here. At Kaldjik, one hour east of Karamanli, Mr. Ramsay and I separated finally. My road led north-eastward down the valley of the Gebren Tchai, a district blank on the map, but containing a number of villages. About two hours east of the village of Eïnesh is the site of an ancient town, now wholly deserted. The remains are not unworthy of notice. Among other things may be mentioned the tombs, most of which are circular buildings of stones hewn into a circular shape, with massive stone foundations. This may possibly be the site of Themisonium. Buldur and Baris, now Isparta, were visited; then Seleucia Sidera, now Egerdir, at the southern end of the lake which bears its name.

At Egerdir I was shown two old Byzantine steelyards. The four sides of the bronze beams were all different, each side being apparently intended for a different standard of weight. The great intervals corresponding to our one, two, three, etc., pound notches, were marked by letters of the Greek alphabet. The heavy weight was a head of Zeus in bronze, filled with lead. The owner demanded thirty pounds for the two, which put them out of my reach. I was anxious to get at least an accurate drawing or copy of the weight-slots or notches ; but the suspicious Turk feared that the value of his property would thereby be diminished, and refused to allow me to make any notes or take any drawings whatever.

A glance at Kiepert's great map of Asia Minor (1852–55) will show that the water from the little Godeh Göl is made to flow northward and empty into the Egerdir Göl. This is copied in the French map published by Kiepert in 1884. The reverse of this is true, as was pointed out by Hamilton nearly fifty years ago.* The outlet of the Egerdir lake is to be found forty minutes south-east of Egerdir ; it is a strong, deep, and very rapid stream, spanned by a bridge just at its exit from the lake. I made inquiries again and again concerning the course of the river, and found the natives unanimous in the statement that the water goes southward to Adalia. They did not tell me, however, as they did Hamilton, that the water from the Godeh Göl finds an underground exit. I am decidedly of the opinion that the Egerdir lake is the real source of the river Cestrus ; but this point will have to be decided definitely on a future journey, and I regret that it was out of my power to trace the matter up at once. The scenery of the Egerdir lake is among the most picturesque and beautiful I recollect to have seen in Asia Minor. Rugged, jagged, threatening mountains, many with snow-capped peaks, spring up almost perpendicularly from the lake on all sides.

From Egerdir the road goes around the lake to the south, over a path which is still frightful, notwithstanding recent attempts to make it passable. But after the so-called Iron Gate (Demir Kapu) is passed, the road is level as far as Gelendus (no ruins !), whence I went by zigzags to Antiochia Pisidiae. Besides the interest attaching to Antiochia of Pisidia as the scene of some of the labors of the Apostles Paul and Barnabas, it is very rich epigraphically. I copied very many inscriptions here, more than half of which are in Latin,

----

* Hamilton, *Asia Minor*, I. p. 482.

testifying to a very large and wealthy Roman colony. Although there can be no doubt in regard to the site of Antiochia, still documentary evidence is by no means abundant, and the following official document may not be out of place here : —

No. 4.

ΑΥΡΔΙΟΝΥΣΙ
ΟΝΤΟΝΑΞΙΟ
ΛΟΓΩΤΑΤΟΝΕ
ΚΑΤΟΝΤΑΡΧΟΝ
ΡΕΓΕΩΝΑΡΙΟΝ
ΗΛΑΜΠΡΑΤΩΝΑΝ
ΤΙΟΧΕΩΝΜΗΤΡΟ
ΠΟΛΙΣΕΠΕΙΚΙΑΣ
ΤΕΚΛΙΤΗΕΕΙΡΗ
ΝΗΣΕΝΕΚΑ*

A few of my inscriptions from Antiochia are already known, but the majority are new. The inscriptions copied up to this point number one hundred and fifty. The ruins, both on the Acropolis of Antiochia and elsewhere in the neighborhood, are very considerable and impressive ; but, as they have been described by Arundell and Hamilton, I need not delay longer over them.

The road (six hours) from Antiochia to Philomelium, now Ak Sheher, leads across the Sultan Dagh by what, in the absence of accurate information, has hitherto been thought to be a pass. But it is a pass only in so far as deep gorges lead up to the great backbone of the mountain on either side. The mountain sends off ridges without number at right angles to the mountain chain, and any two opposite gorges may be called a pass with as much propriety as the one through which the road from Antiochia to Philomelium leads. There are some high peaks in Sultan Dagh at the north : but the point at which the road crosses is quite as high as any other in the mountain, and both ascent and descent are very tortuous and laborious. The road reaches the great plain of Philomelium one hour north-west of that city, and consequently does not descend the gorge at the mouth

---

* The chisel of the engraver will make small slips sometimes!

of which Philomelium lies, as it appears to do on Kiepert's map.
But few remains of Greek antiquity are to be found at Philomelium ;
but on the other hand the traveller is surprised by some Seldjuk
ruins of exquisite beauty. The accurate workmanship displayed,
even in the execution of details, will compare favorably with Greek
buildings of a good period.

At Philomelium I was joined, as had been previously arranged, by
my friend, Mr. J. H. Haynes, of Robert College, Constantinople,
who accompanied me as photographer during the rest of the journey.
Thanks to his art, we have photographs of the Seldjuk ruins of
Philomelium. My travelling-outfit had been left at Smyrna, so that
I had had a hard journey thus far. Mr. Haynes's advent was therefore hailed with delight ; for henceforward we could have substantial
food, on which depends in great measure the success of an expedition like this. Four more horses were bought here, one for Mr.
Haynes, and three for the photographic plates and other baggage.

From Philomelium my route lay along the foot of Sultan Dagh in
a south-easterly direction to Doghan Hissar. This region is very
populous, and what is a blank mountainous space on the map is in
reality a plain full of prosperous villages. These villages, as Doghan
Hissar is approached, have numerous inscriptions, mostly late. At
Kara Agha I copied one which proves that Hadrianopolis Phrygiae
was somewhere in this neighborhood.

No. 5.

```
AYPHΛЄIOCZѠ
TIKOCΠAYΛЄINOY
AΔ░░░NOΠOΛЄITH
CTH░CYNBIѠAY
PH░░░IΔAΓΛYKYTA
THMNHMHCXAPIN
```

In point of fact, Mr. Haynes, who had taken a different route from
myself, found ruins at Reghiz, but especially at Kotchash, one and a
half hours north-east of Doghan Hissar. These ruins are late and the
inscriptions are Byzantine ; but still Kotchash is probably the site of
Hadrianopolis. Doghan Hissar is a modern town, without any
antique remains. From Doghan Hissar our road lay westward, in

the direction of Kara Agatsh, it being my object to investigate the pass of Sultan Dagh, in the hope of finding an ancient town (possibly Pappa). The road enters the mountains from Kara Agha; the ascent is gentle but steady. The descent on the west side of the mountain is sharper and more precipitous. The pass is low, and no trace of a town was found. It will be noticed that the west side of Sultan Dagh is a blank on the map; but the district is densely populated. In a fertile valley about an hour east of Kara Agatsh there is a cluster of seven large and prosperous villages. The whole community goes by the name of Tcharük Serai; but each of the seven villages has its own distinctive name, and each of these names has the ending *mahalli* (instead of *kieui*), *e.g.*, Tchikourmahalli, Ulumahalli, Sugharmahalli, Belükmahalli. As a case parallel to this may be cited Yalowadj, which is composed of five villages in a cluster, each with the above ending.

Tcharük Serai is certainly the site of an old town; possibly Pappa (or Amblada) must be placed here, but no documentary proof exists at present. That Phrygian was the language of the aborigines is clear from the following Phrygian epitaph: —

No. 6.

IOCNICEMONKNOYMA
NEKAKONΔAKETAINI
MANKATIETITTETI
KMENOCEITOY

Another Phrygian fragment, from the Mosque of Aïplar (one hour south of Kara Agatsh), may be inserted here.

No. 7.

IOCKECEMONTOKAKONOL

In the summer of 1883, Mr. Ramsay and I found several Phrygian inscriptions in the plain east of Sultan Dagh, at Arküt Khan and Ilgün; but, so far as I know, these are the only ones known on the west side of the mountain.

From Tcharük Serai we returned to Antiochia, visiting the many villages, copying inscriptions, and making route surveys. At Managha,

about four hours south-east of Antiochia, I found the fifth milliarium from Antiochia. It has therefore been transported about
seven miles from its original place. The stone is badly defaced on
both sides.

No. 8.

First Side.

DDNN
FLCONSTANTINOMAX
▨▨▨TAVG▨▨▨▨
FLIVLCONSTANTIOET
CLCONSTANTI▨▨▨
VICTORI⸮⸮≶⸮MPAVGG
ABANTIOCHIA▨▨▨
▨▨▨▨▨▨▨▨▨▨▨▨▨▨
PONTIF·MAX·TRIB
POTXIIICOSIII
P·P

M P U

No. 9.

Second Side.

IMPCMAVRVAL▨▨
MAXIMIANO▨▨
FINVICTAVG▨▨
▨▨▨SONΓPA▨P▨
R▨▨▨▨▨▨▨
▨▨▨CAESARIB▨▨

[*Uncut space.*]

▨▨▨IMPCAS▨▨
MACAPO▨▨▨
ETIMFCNEV▨▨
IMORVAL▨▨▨
MAXIMINO▨▨
REAVGⱯ▨▨

I regret that I could not spend more time in this district. It did not fall within the limits of my original plan, and my visit was necessarily a hasty one. There are several points that will need investigation on a future journey. For instance, at Karakuyn, a *kale* in the mountains two hours south-east of Antiochia, extensive ruins and inscriptions were reported to me by the people of Yalowadj. Karakuyn is almost certainly Oroanda or Misthia. In this connection may be cited an inscription which throws a valuable side light on the geography of Pappa and Oroanda. The inscription was found at Hissar, half an hour east of Antiochia.

No. 10.

TYXHNEY
MENHTH
KOΛWNEI
ATIBEPIO

ΠΟΛΕΙΤWΝΠΑΠ
HNWNOPONΔE
WNBOYΛHΔHMOC

The KOΛWNEIATIBEPIOΠΟΛΕITWN is Antiochia.

Ruins and inscriptions were also reported at Bachtiar, four or five hours south of Antiochia, at the foot of Sorkundja Dagh. In this region Neapolis must be looked for.

From Antiochia we returned to Kara Agatsh, a large town situated in the centre of a very fertile plain. Anabura, a town mentioned by Livy in his account of the march of the consul Cn. Manlius, has always been placed far north of Kara Agatsh and Antiochia, and, as I think, correctly. At Kara Agatsh I found the following inscription : —

No. 11.

OBPIMIAN
OCKAIM⊗
CAIOCOIIOY
ΛΙΟΥΤΟΠΡ
5   ΟΓΟΝΙΚΟΝ
ΕΡΓΑCΤΗΡΙ

```
          ΟΝΚΑΤΑΓΑΙ
          ΟΝΥΠΟΒΑΛ
          ΟΝΤΕϹΤΑϹ
   10     ΠΑΡΑϹΤΑΔΑ
          ϹΚΑΙΤΗΝΟΡ
          ΟΦΗΝΚΑΙΤΟ
          ΗΝШΜΕΝΟΝ
          ΑΥΤШϹΥϹΤΡ
   15     ШΜΑΠΟΔШΝ
          [uncut] ϹΥϹΤΡШ
          ϹΑΝΤΕϹΚΑΙ
          ΤΑΛΟΙΠΑΠΑ
          ΝΤΑΚΟϹΜΗϹΑ
   20     ΝΤΕϹΑΥΤΟΥ
          ΕΚΤШΝΙΔΙШΝ
          ΑΝΑΒΟΥΡΕΥϹΙ
          ΝΕΠΟΙΗϹΑΝΕ
          ͞ΖΕΔΡΑΝΟΝΤΕ
   25     ϹΑΠΟΓΟΝΟΙΜΑ
          ΝΟΥΟΥΡΑΜΜΟΟΥ
```

For the present I call attention only to lines 22 and 23. Besides this inscription, I was informed that ruins and inscriptions were to be found at a place called Enevre, said to be from two to four hours south-east of Kara Agatsh. "Enevre" is almost certainly a corruption of "Anabura," and the inscription and the name of the village seem to point to the fact that at least *an* Anabura once existed in this neighborhood. But it is certain that it is not the place touched by the consul Manlius. It is not uncommon, either in Asia Minor or Greece, to find two towns or rivers bearing the same name.

Hence our route led south-east towards Carallia, whose name is still preserved in the Turkish *Kürili*. It must be noted, however, that the old town of Carallia lay an hour north-east of Kürili. The Acropolis is still easily identified, and architectural fragments are found in a cemetery near a large spring north-east of the Acropolis. The inscriptions of Carallia are all Christian.

Our next point was Elflatoun Bounar, which we visited to photograph the sculptures mentioned by Hamilton. These sculptures, being known only by hearsay, will be welcomed by archæologists. The very ancient structure bearing the sculptures is just on the edge of an immense spring; in fact the whole region abounds in large springs of delicious cold water. The water from the great valley of Afium Kara Hissar, Ak Sheher, and Ladik has no visible outlet, and must find an underground exit. No doubt part of it flows out in these springs, and part by the many great springs at Tchifteler, Orcistus, Amorium, Abrostola, etc., in the woodless country north-east of the valley.

Hence we went to Selki, the Serki Serai of Kiepert's map, which places it much too far to the north. The region north-west of Selki is full of villages, which lie along the foot of Sultan Dagh. At one of these villages a large spring of hot water is reported, which is said to be a popular resort of the Turks of this region. From Selki the road passes through a wild mountainous district, infested by brigands, to Kizil Ören. About half an hour west of Kizil Ören there still stand a Seldjuk khan and mosque in a fair state of preservation, but the remains of a remoter antiquity are all Christian.

Iconium is eight hours distant from Kizil Ören; the road is uninhabited and monotonous. About two hours from Kizil Ören there is a Seldjuk khan, not well preserved. A second khan is five hours distant, at the junction of our road with the road from Iconium to Philomelium.

At Iconium I found quite a number of inscriptions, most of which are late and of little value. The people of this eastern country seem to have had little interest in the affairs of this world, and spent their surplus energy in preparing tombs and epitaphs for themselves. When Leake passed through Iconium, the walls of the town were full of inscriptions, which he had no time to copy. After the destruction of Iconium by Mehemet Ali of Egypt, these walls were used as quarries for the buildings of the modern city of Koniah. The inscriptions mentioned by Leake all perished in this way before an epigraphist was found to copy them. But many inscriptions are no doubt still in the walls with the inscribed side hidden from view. Part of the wall had been thrown down only a short time previous to our visit, and I copied several inscriptions brought to light in this way. The walls were built in the common Greek fashion (Thuc. I. 93); that is, two walls were built at a fixed distance apart, and the space between them was filled with earth and stone débris. At Iconium the filling consisted mostly of simple clay or mud, which took faithful impressions of the stones composing the outer shell of the wall, so that one may now see therein neat reliefs of inscriptions, Phrygian doors, and architectural fragments. The ruins of the buildings erected by the early Seldjuk Sultans of Iconium, from Aladdin down, are, for the most part, of exquisite beauty. Mr. Haynes spent two days in photographing them; and as very few travellers go to Iconium, these photographs will no doubt be acceptable to many.

The Governor of the Vilayet of Koniah, Sahib Pasha, who studied in England and speaks English fluently, showed us kind attentions in more ways than one. He is collecting the most important antiquities of the district, as they come to light, for the Museum in Constantinople, and his collection is not without interest. Among other things may be mentioned a frieze in very high relief. Unfortunately we were unable to get photographs of the collection.

The road from Iconium to Archelaïs, now Ak Serai, crosses the desert region. The first station is Obrukli, the ancient Savatra, at a distance of fourteen hours. We found no water on this journey. The plain is absolutely level, and the thirsty traveller is mocked on all sides by the *Fata Morgana*, promising water near at hand; but the promised water recedes continually, and finally turns out to be nothing but a deceptive mirage. At Obrukli there is a little lake, the surface of which is about ninety feet below the surrounding country. The villagers use this water for household purposes. We were told that the water is drinkable at all seasons of the year, except for two weeks in December, when it is in a state of violent ebullition. When this season approaches, they lay in a sufficient supply of water to last until the lake has resumed its wonted calm. How true this may be, or what causes the phenomenon, I am not prepared to say.

Sultan Khan, the next station, is the grandest and most beautiful of all the remains of Seldjuk splendor seen by us in Asia Minor. We spent one day in its welcome shade, during which time numerous photographs were taken, and the huge building was roughly measured. One of the Arabic inscriptions states that it was built A.D. 1277. A very large spring rises quite near to Sultan Khan, and the land yields abundant harvests wherever it can be properly irrigated. Indeed, this is true everywhere in Asia Minor.

Archelaïs is a sleepy, uninteresting town, with but few traces of the Græco-Roman civilization; but the foot-prints of the Seldjuks are abundant.

At Selme, three and a half hours east of Archelaïs, we found numerous dwellings cut in the rock, similar to those described by the early travellers at Soghanli Deressi and Udjessar. In fact, we found these wherever the soft volcanic tufa appears (Hamilton, I. 97). Selme is situated in a deep gorge through which the Irmak flows, and in which, in fact, it has its source. The cliff to the east rises perpendicularly from three to four hundred feet; at its base there is a maze of sharp natural cones, similar to those at Udjessar. Most of these cones are hollowed out, often with several stories, for human dwellings, and are used as such now, as in ancient times. The whole cliff is honeycombed into dwellings, chambers, chapels, passages, and tombs; story rises upon story. Even now, people live and die in these rock-cut dwellings, at least two hundred feet high on the cliff. There is no earthly reason why they should live there, as the country is safe and land is abundant; but they do not seem to object to the dark winding stairs and passages.

Across the Irmak, five minutes south of Selme, is the village of Ichlara, the cliff behind which is also similarly honeycombed; several façades of temples are conspicuous on the side of the cliff. A short distance east of Ichlara the Irmak gushes out at the foot of the cliffs, a full-grown river at its source.

Our road hence led by way of Kuyulu Tatlar, so called from the numerous wells which supply the village with water, to Melegobi. This region, though blank on the map, is full of villages, most of which will appear on the next map. It may be noted that the Tada Su of the map does not exist, at least not in the plain of Kuyulu Tatlar and Melegobi; and, furthermore, the drain-water from this district must run south, and not north as on the map. Melegobi is a large and flourishing village, inhabited almost exclusively by Greek-speaking Greeks. The Greeks are numerous all through the western part of Cappadocia, and generally cling to their language with great tenacity, a fact worthy of notice, inasmuch as the Greeks in other parts of Asia Minor speak only Turkish. Instances of Greek-speaking towns are Nigde, Gelvere, Melegobi (Μελεκόπια), and Ortakieui in Soghanli Deressi.

Hence we travelled to Soghanli Deressi, the wonders of which have been described by Hamilton. The rock-cut dwellings are more numerous, but of the same character as those at Selme and Ichlara; only at Soghanli Deressi there are no temple façades to be seen. Soghanli Deressi is simply a break in the surrounding plateau from three to four hundred feet deep. The descent from the plateau to the valley is made by a very steep road hewn out of the volcanic tufa. At its head the valley of Ortakieui is about one hundred yards wide; but the width increases steadily, and it is from five to seven hundred yards wide at the point where Soghanli Deressi branches off laterally from it. While the surrounding plateau is a barren waste, the soil in the valley is exceedingly fertile, delighting the eye with its luxuriant gardens.

Whether these rock-cut habitations date originally from an earlier epoch or not, it is at all events certain that they were used by the early Christians. Chapels are numerous, in some of which may still be seen pictures of Byzantine Saints with inscriptions just like those common in orthodox churches of to-day. Among the saints depicted are Σέργιος, Βάχος, Μερκούριος, names which may give a clue to the time when Christians worshipped here. In the floor of the chapels graves were cut, in some of which we found human skeletons. Indeed, such tombs are frequent in the dwellings themselves, so that, as Hamilton remarks, the people lived in the same room with their pigeons and their dead. We have a goodly number of photographs from Soghanli Deressi.

Zengibar Kalessi is situated about half an hour west of Develi Kara Hissar. It is a lofty rock with two peaks, one of which is considerably higher than the other. In the saddle between the two peaks nestles Kalekieui. There can scarcely be a doubt but that the higher peak of Zengibar Kalessi is Nova, the proud rock where Eumenes and his little band defied the whole army of Antigonus for nearly a year. From Develi Kara Hissar we pushed on northward to Indjesu, and through the Sazlük, or *place of the bulrushes*, to Caesarea, thus passing almost half around the snow-capped peak of Mt. Argaeus. It was my intention to go directly from Develi Kara Hissar to the Antitaurus, not touching at Caesarea until the home journey; but circumstances made a visit at this time imperatively necessary. We thus made a great detour, and lost four or five days, all on account of one packsaddle ! At any rate, we learned the important lesson that every article in the outfit for a journey of this kind must be of the very best quality.

Parting with regret from our kind friends, the American missionaries of Caesarea, we hurried past Tomarza to the Antitaurus, which was crossed by the precipitous pass between Dede Dagh and Beyli Dagh. This region was hitherto unknown ; we found it fertile and well populated, and of course route surveys were made here, as on the whole journey. We visited Comana, the only place marked on the map, about two hours south of Olakaya. The Great Goddess is no longer worshipped in Comana ; but, to our immense astonishment, we found a Protestant church there, composed of the converts of the American missionaries. From Comana, now Shahr Deressi, we took a nine hours' journey down the river Sarus to Hadjiu, which is also a seat of the American missionaries. It is on the right side of the river, and about three hours distant from it. Hadjiu is a modern town, inhabited solely by Armenians, and is situated in a great hole in the mountains about 1500 feet below the level of the surrounding country. Three hours north-east from Hadjiu, the great cañon of the Sarus is reached. The cañon is fully 1000 feet deep. The banks are almost perpendicular, so that one can scarcely believe it possible for a living being to descend and ascend ; yet it may be done. Five hours south-east of Cocussus, at an Avshar Yaïla, known as Kilissedjik, we found two Greek tombs of a good period. I am inclined to place Laranda here.

The plain of Cocussus is remarkable both for its exuberant fertility and for its springs and rivers. A dated inscription of the ninth year of Trajan (107 A.D.), found at Deghirmeri Deressi, informs us that Zeus Epikarpios was then worshipped here ; indeed, in so fertile a plain we should naturally expect to meet with the cult of some god of the harvest.

No. 12.

ΕΠΙΝΕΡΟΥΑΤΡΑΙΑ
ΝΟΥΚΑΙCΑΡΟCCΕ
ΒΑCΤΟΥΓΕΡΜΑΝΙ
ΚΟΥΔΑΚΙΚΟΥΕΤΘ
ΔΙΙΕΠΙΚΑΡΠΙШ
ΚΑΠΙΤШΝΤΙΛ
ΛΕΥCΕΚΤШΝΙΔΙШ
ΝΑΝΕΘΗΚΕΝ

Tasholuk, a village one hour south of Cocussus, is the site of an old town.

Among a goodly number of inscriptions copied at Cocussus were several Roman milliaria. I have been told that the milliaria found by me in the *terra incognita* of Antitaurus are of so great importance that I have no right to keep them from the public until my return from Babylonia. In this Preliminary Report I cannot do more than give the inscriptions in uncial text, hoping to make historical and explanatory remarks upon them at some future time.

These milliaria are about eight feet high, and from two and a half to three feet in diameter at the bottom, tapering off to a very thick, blunt point at the top. They are accordingly cone-like in shape. The stones are very rough and unpolished, and the surface is full of elevations and indentations. It is obvious that inscriptions on such a rugged, unequal surface are very difficult to read, and that, without some practical experience in epigraphy, one would stand before them absolutely helpless. Impressions of such inscriptions are altogether worthless, as trial has proved to me conclusively.

I.

MILLIARIA AT COCUSSUS (GÖKSÜN).

No. 13.

Göksün.   Western Cemetery.   Waddington's 33.

IMP
CAES
DIVISEVERNEP
DIVIMANTONINI
FIL·
MAVRANTONINO
PIOFELICIAVG
MILIARESTITVTA
PERMVLPOFELLI
VMTHEODORVM
LEG AVG
PR   PR

ΡΛΓ

One hundred and thirty-third milestone.

No. 14.

Göksün. Western Cemetery.

```
S A L V A L
X I M I A N O
L V I C A E S
```

```
A N T O N I V S G O R D I A
N V▨▨O B I L L S I M V S
E X A R R E S T I T V I T
P E R C V S P I D I
A M I N I V M S E
V E R V M L E G E T P R
P R E T O R E M
```

P M A

One hundred and forty-first milestone.

No. 15.

Göksün. Southern Cemetery.

```
I M P
C A E S A R C I V L
V E R V S M A X I M I N V
▨▨▨C A E S S N ^▨
G A I O ▨ I A▨▨L I▨
D L▨▨L E I I A N O
E T I N V I C T O▨ ^ V▨
N O B I L I S S I M V S C A E S A R
V I A S E T P O N T E S V E T V▨
T A T E C O N L A B S A S R E S
T I T V E R V I▨T▨▨
P E R▨▨▨▨▨
▨▨▨▨▨L E G▨▨
A V G G   P R   P R
```

XII # P M A

One hundred and forty-first milestone.

No. 16.

Göksün.   Southern Cemetery.

```
IMPCAESAR
DIVISEVERI
NEPDIVIM
A   ITONINI FIL
MAVRANTON
NOI
         FE
LICIAVG
       SIS
MILIARE
STITVTAPER
MVLPOFEL
LIVMTH
    O
```

No. 17.

Göksün.   Southern Cemetery.   Badly worn and illegible, except

```
LEG
PR  PR
```

P Λ

One hundred and thirtieth milestone.

No. 18.

Göksün.   Southern Cemetery.

```
IMP


AVR
RIB
COSSAPP
TESVETTVSTA
NLAPSASRESTITV
ITΔPKE
```

There are several uninscribed milliaria in the cemeteries of Göksün.

From Göksün we went back in a north-west direction to about six miles from Comana. On this excursion also we found a number of milliaria, and thus were enabled to trace the Roman road from Comana to Cocussus in its entire length.

II.

MILLIARIA OF THE ROAD FROM COMANA TO COCUSSUS.

Mehemet Bei Kieui. One hour north-west of Göksün.

No. 19.

```
        I M P C C C
Y       D I O C L E T I A N O
  I          P ⌐ I ∪ \ I I
        A      C
                    T I T
                    M P
                      I E
```

No. 20.

Mehemet Bei Kieui.

```
    I M P C A E S
    A R M A R C V▨
  ▨V L P H I L I P P V S
   F E L I X I N V I C T V S
  ▨V G E T M A R C V S
  ▨H I L I P P V S N O B I L I S S I
  ▨V S C A E S A R  V I A▨
  ▨P O N T E S  V E T V▨
  ▨E C O N L A P S A S R▨
   S T I T V E R N 7 P E R▨
  ▨O N M M E M M I V M H
   ▨E T F V A L▨
   ▨C O N S T A N T▨
   ▨▨N O B C A▨▨
      ▨▨S ( ▨▨
```

No numerals remain. There are two other uninscribed stones at this place. Half an hour south of Kekli Oghlou is a stone almost entirely buried.

No. 21.

Kekli Oghlou.   Four hours north of Göksün.

```
        C A E
      A R M A R C V▨
      P H I L I P P V S  P I V S  F▨
      N V I C T V S A V G
    ▨＼R C V S I V L  P H I L I P P▨
      B I L I S S I M V S  C A E S
      A S E T P O N T E S  V E T▨
    ▨＼T E C O N L A P S A S  R E S▨
      R A P E R  A N T O N▨V▨
      M I V M  H I E R O N E M
    ▨E G A V G G  P R
      P R
```

No. 22.

Kekli Oghlou.

```
      I M P C A
      E S A R I G A
      I O I V I I ° V E
      R O M A
        [Uncut space]
      M I N O▨P I O
      F E L I C I▨A V G
      T R I B▨P▨Γ E
      L I C I N N I V M
      S E S E  I M I A N
      V I▨ L E G▨A V G
      P R  P R
```

РΛН

One hundred and thirty-eighth milestone.
Also, two uninscribed stones are at Kekli Oghlou.

No. 23.

Yalak. Two hours from Comana.

```
ARC
LIPPVS
SSIMVS
SARVIASETP
ONTESVETV
STATECONL
PSAS R   ST   I
ERVN
NIVX
IVM
 MVC
  M
```

No. 24.

Yalak.

```
CΛЄSA
V
OC
 VIDΛE
  OV
 LCISΛ
 PON
ONLAPSAS
```

No. 25.

Yalak. The only milestone found with a Greek inscription.

```
   CIACYΠATO
   OC TAC OΔOYC
   TOIOI    OY
        NTICT
```

P M Δ

One hundred and forty-fourth milestone (see below, p. 35).

The *Antonine Itinerary* for the whole Antitaurian region seems hopelessly confused, and its inconsistencies will perhaps never be satisfactorily explained. On p. 210 we read : —

A Coduzalaba	
Comana	XXVI.
Siricis	XXIIII.

while on p. 211 we have the following : —

Item a Caesarea Anazarbo CCXI., sic:

Arassaxa	XXIIII.
Coduzalaba	XXIIII.
Comana	XXIIII.
Siricis	XVI.
Cocuso	XXV.

Now the milliaria given above show that the Roman road went, as one would naturally expect, by Mehemet Beikieui, Kekli Oghlou, and Yalak ; and as the whole distance between Comana and Cocussus is reckoned as eight hours, there is plainly something wrong in the statement of the *Antonine Itinerary*. Both Kekli Oghlou and Yalak are sites of small ancient towns; but the most important of these was at Yalak, and at Yalak I am inclined to place Siricae. In that case the *Antonine Itinerary* would be nearer the truth if it were emended to read : —

Comana	XXIIII.
Siricis	VI.
Cocuso	XV.

Let it be noted that this, besides being a direct route, is the only natural road from Comana to Cocussus : on the north lies the Biu Bogha Dagh, and on the south the Yuvadja Dagh. It is unreasonable to suppose that the Romans would neglect the only natural roadbed to carry a road over the huge mountains just mentioned.

On our return journey to Cocussus we followed the Tölbüzek Su to its source, which is about three-quarters of an hour west of Mehemet Beikieui, at the foot of Yuvadja Dagh. Here innumerable springs gush from the mountain side, and the water from them is sufficient to form a large and swift river of the purest, coldest water.

From Cocussus we turned our faces eastwards in the direction of Arabissus. The present road, to all intents and purposes, follows the ancient Roman road, most of the milliaria of which were found by us.

III.

MILLIARIA ON THE ROAD FROM COCUSSUS TO ARABISSUS.

No. 26.

At an old cemetery by the roadside, forty minutes east of Göksün.

```
R V S  ·
A R A O I  A R
P O T I V I I
T I M P C A E S
///////////////////
R E S T I T V E R V N T
A N V M  L E G  P R P R
```

In this cemetery there are four milliaria, one of which is deeply buried.

No. 27.

Ibidem.

```
M A X I M I A N O
N O b  C A E
S S
```

No. 28.

Ibidem.

```
I M P
///M A V///
M///P E P///
C///////M A X I M///
C O A N T O R///////
C O R L N O C C A E///
L I C I  A V G///T O
R E S T I T///S R V N T
P E R  C V S P I Δ I M /
M I N I u M  S E V E R V M
C A T V M  P O P R A C///
T O A///////
```

In a cemetery one hour and five minutes east of Göksün there are two more milliaria; one nearly buried, the other erect but illegible.

It was impossible for us to get at the half-buried stones : to raise one out of a hole is generally half a day's work for four men, in a country where levers are not to be had.

No. 29.

In a cemetery one hour and forty minutes east of Göksün.

PER MEMM

No. 30.

Ibidem. Erect.

T₁ MAXI ONTIM
M XII COS IIIiP IBO
IMI AVBE ANTONINYS
I T
PERHYLIYMFLACICYMIAEWAYM EO

No. 31.

Ibidem. Erect.

I M
LSE
PIVS
PARTI
IMPXI
ⅡAVR⁻
ETLISE
PERCIVL

No. 32.

Ibidem. Erect.

dIOCLETIA \
E T ∧ N
ITCA IV
CONSTA TIO
ETCAIVM
MAXIMIANO
N PR

No. 33.

Ibidem. Erect.

AES
C/ OIVL
ROMAXIMI
PIO FELICIA
VG TRIBPPERLI
CINNIVM SERENI
ANVN LEG AVG
PR PR

P K B

One hundred and twenty-second milestone.

In the cemetery of Kanlükavak we found no less than twenty-six milliaria, many of which were never inscribed. The inscribed stones cost us a day and a half of hard work, in deciphering and copying the inscriptions (Nos. 34–48).

No. 34.

Kanlükavak. Cemetery.

MP
SPI
ICTV
ARCVS
NOBILISSIM
_SARVIAS ET TO
T S VETVSTATE
CONLAPSAS RESTITVE
PER ANTONIVM M C
MIVM HIERONEM
LEG AVG
PR PR

No. 35.
Ibidem.

V
M
Є
ICI
VNI
O N
PR PR

P K

One hundred and twentieth milestone.

No. 36.
Ibidem.

\Nᵢ　　　）NO
LISSIMOCASA
CATCLEMЄNT
CRᴄRCROUII ᴠCIᴧ
IMP

P K Є

Possibly the one hundred and twenty-fifth milestone.

No. 37.
Ibidem.

IMP
DIVISEVERI
NEPDIVI MAN
TONINI FIL
Mᴧ VR
NO PIO FELICI
AVG
MILIARESTITVTA
M▨▨▨POFELLIVM
THEODORVM
LEG AVG PR PR

M ▨ K I

Possibly the hundred and twenty-first milestone.

No. 38.

Ibidem.

PR PR

-

No. 39.

Ibidem.

I M P
C A E S A R I M A
R C O Λ N T O N I
O G O R Δ I A N O P I
O F E L I C I A V G V S
T O R E S T I T V E R V
N T P E R C V S P I Δ
I V M F S A M I N I
V M S E V E R V M
L E G A T V M P R O P
R Λ Є Ъ T O R E M

No. 40.

Ibidem. Two inscriptions are so inscribed on and over each other
that it is perhaps impossible to untangle them; but the following
can be read : —

I M
C A E S M A R C V S
˙I V L P H I L I P P V S
P I V S F E L I X

No. 41.

Ibidem.

I M P P
O C░T I A I
Є T I˥ A L T I U A L
M A X I M I A N O
P P F F I N V A V G
E T F I Λ U I U Δ I
C O N S T A N T I O

EIC ^ I AI
CAES
MAXIMIANO
[*uncut space*]
H MAXPONTMAXTRIBPOTVIO
IP XI COSII PP PROCOS ET IMPCAES
MAVREL · ANTONI▨▨▨▨
NVSAVG [*name erased*]
[*erased*]   TITVERVNT
PER C· IVLIVM FLAC
CVM AELIANVM LEG PR PR

No. 42.

Ibidem.

IMPP
dIOCLETIANO
IMP CAE^TS^MAVRVAL
MAXIMIANO
LSEPTIMIVS SEVERVSAVG
PPLE I NV
PIVS PEPTINAX AVG ARABIADIAB
PARTH MAXPONT MAXTRIBTiiOTVI
OICT
IMPXI COSIIIPPPROCOS ET IMPCAES
ET CAIVA
MAVREL ANTONINVS AVGNO
ETLSEPT▨▨VS NObb CAESS▨ESTITVERVNT
PERC· IVLIVM FLACCVM AELIANVM LEG PR PR

Nos. 43 and 44.

Ibidem.

I M
R CAESMA
RCVS IVL P
HILIPPVS PIV
S EELIXINVI
CTVSAVG E
T MARCVS

```
I V L I V S  P H I
L I P P V S  N O
B I L I S S I M V S
C A E S A R  V I A
S  ET PONTES
▨' E T V S T A T
▨▨ N L A P S A S
▨▨▨▨⁻ V E R
▨▨▨▨▨ᕲ A
```

On the other side of the same stone there is this : —

```
I M P P E Ɛ C ᴎ A ɩ
ᕷ I O C L E T I A N O
E T  M A V R V A L
M A X I M I A N O
Ϝ P F F  I N V I A V G
V I  V A I
C O N S T A N T I O
E T  C A I ▨▨ A ▨ E
M A X I M I A N O
N O b b  C A E S S
```

No. 45.

Ibidem.

```
A X
P P  F F I N V I  A V ᕼ
E T  F L A V I  V A L
C O N S T A N T I O
E T  C V A L E N
M A X I M I A N O
N O b b  C A E S S
P
```

Nos. 46, 47, 48.

Ibidem. Stone with three inscriptions inscribed on and over each other. After much labor I succeeded in disentangling them.

No. I. reads : —

IMPCA
ESAR GVI'IVSTREBO
NGALLVS ET IMPCAES
ARGVIVIVS VELDVMI
NIANVS VOLVSIANV
PII FELICINVICTIAVGGVIAS
ET PONTES VETVSTATE CON
LAPSAS RESTITVERVNT PERA
VERGILIVM MAXIMVMVC
VGG PR PR

No. II. reads : —

RESTITVTA
////ERMVLP
OFELLIVM
THEODORV
MLEG AVG
PR PR

M

No. III. reads : —

IMPPCC
dIOCLETIANO
ET MAVRULI
MAXIMIANO
PPFFINVI AVG
ET FIAVGI VAL
CONSTANTIO
ETCAI VAL
MAXIMIANO
NObbCAESS

Between Kanlükavak and Arabissus no milliaria were found. Indeed all seem to have been transported from this whole region to the cemetery of Kanlükavak to serve as tombstones. Arabissus, now Yarpuz, was once an important place, to judge by the remains still extant, which, however, are mostly Christian.

About one and a half hours west of Arabissus is Ziaret Serai, a Seldjukian palace or villa. In a cemetery midway between Emirli and Ziaret Serai I found the one hundredth milestone (No. 49).

No. 49.

Cemetery between Emirli and Ziaret Serai.

MPTRIBPOT VIA
ET////NTE
SR///STITVERVN
CIVLI////OCI
AVG PR

C

All the rest of the stones have Greek numerals, this one alone having the Latin C.

At Yalak the one hundred and forty-fourth milestone (No. 25) was found. It will be seen that the numerals diminish steadily along the road from Comana to Cocussus, Kanlükavak, Arabissus, a fact which proves conclusively that Mr. Ramsay was correct in his opinion that distances in the Trans-Antitaurian region were measured from Melitene.

From Arabissus we undertook a zigzag journey in a northerly direction, with Khurman Kalessi as an objective point. This excursion was disastrous to man and beast, and its results were purely chorographic. Our return journey led us by way of Tanir, situated on the Khurman Su just at the point where it emerges from the mountains. The traces of antiquity are abundant at Tanir, and the name would seem to indicate that it is the site of Ptandaris. Hence we returned to Arabissus.

No. 50.

Cemetery of Yarpuz. So super-inscribed as to be hopelessly illegible.

NOBILISSIM \
CAES

No. 51.

Cemetery of Yarpuz.   Erect but illegible.

CONLAP

IV.

MILLIARIA ON THE ROAD FROM ARABISSUS TO MELITENE.

No. 52.

In Cemetery one hour and four minutes east of Yarpuz.

I
RESTIT
PER
CIVLIVM FLAC
CVMAELIANVMLEGPRPR

MIL P

This is another hundredth milestone, with the numeral in Greek.

No. 53.

Ibidem.

IMP⅗CAESAR
AVREL
[*space overwritten*]
POTEST⅗COS
VIASETPONT
VETTVSTAT
APSAS REST
T.⅗

Two more stones in this same cemetery are wholly illegible.

Nos. 54 and 55.

[These two inscriptions were not received from Dr. Sterrett in time to be printed here. — EDD.]

At Albistan we found nine milliaria, some of which were never inscribed, and the rest, with one exception, are wholly illegible.

No. 56.

Albistan. Cemetery.

CAES
E P
////////////////////
N I ᴗ Λ
////////////////////
RESTITVTA
P OFELLI
VM THEODORVM
AVG PR PR

Owing to illness we were delayed several days at Albistan, during which time kind attentions were showered upon us by the American missionaries.

Henceforward no milliaria were found. I am wholly unable to account for this, as there are only two possible roads from Albistan to Melitene, one of which we traversed on the way out and the other on our return. It may be safely affirmed, however, that the Roman road did not go by way of Köz Agha and Poulah, since this whole road is much too difficult. Had the Roman road gone this way, it could not have avoided the abrupt pass of Ölakaya, and it is exactly this pass which makes it necessary to look for it elsewhere. The only other route is that by way of Derende, and thence down the Tokhma Su, via Argas to Malatia.

From Albistan we visited Arslan Tash, six hours to the north-east, to get photographs of the very antique lions which are there. Thence we returned to Demirdjili, Köz Agha, and across the pass of Ölakaya to Poulah and Melitene.   The chorographic results are considerable. They do not fall within the limits of a paper like the present, and, in fact, can be worked up properly only by a professional geographer. Professor Heinrich Kiepert of Berlin had the goodness to make two special maps for my journey, at the cost of great labor to himself; and in acknowledgment of this I have sent him all my route-surveys and other topographical matter.   He has delayed, as I am informed, the publication of his new map of Asia Minor, in order to insert the routes explored by Mr. Ramsay and me during the past summer.

The new city of Malatia, being the midway-station between Constantinople and Bagdad, is a wide-awake business town, and in this respect it differs very materially from the ordinary Turkish town.   When Mehemet Ali of Egypt was at war with his master, the Sultan, a large number of Turkish troops were quartered for an indefinite period on the people of old Malatia, which stood on the site of Melitene. This was more than the Turks, long-suffering though they are, could bear ; so they abandoned their old homes to the soldiers, and built a new city among the gardens seven or eight miles west of Melitene. Consequently, old Malatia is now a mass of ruins, among which may be seen many fine specimens of the ornamented architecture of the Seldjuk builders.

From old Malatia our road lay through an exceedingly fertile district to the junction of the Tokhma Su with the Euphrates.   Henceforward our faces were turned steadily towards the west, and our homeward road led us by way of Arga, and thence through a very mountainous region, inhabited solely by Kizil Bashi Kurds, an inhospitable, murderous set of filthy villains, who still preserve all the ferocious characteristics of their ancestors, the ancient Καρδοῦχοι, of whom Xenophon has little good to report in the *Anabasis*.   West of Arga some traces of the old Roman road are to be seen, but no milliaria.   At the Beli Gedik we reached the Tokhma Su, and followed it up to Derende, which means " in (or at) the gorge."   An hour east of Derende the gorge is entered, and the new town stretches out along the river for the whole distance between this point and the Derende of the map.   Old Derende was abandoned like Malatia, and for the same reason.   At old Derende the river has cut its way through the rock, which rises perpendicularly about three hundred feet on either side.   The width of the pass through which the river thus flows is about fifty feet. On the right bank is the almost impregnable castle, probably dating from the early Turks ; at its foot, on the west, lies the abandoned town. Midway between Ashta and Arslan Tash we found a very ancient lion in black basalt, and took photographs of it from various positions. From Arslan Tash we took a look at the utterly unknown Palanga Ova, passing via Ketchi Magara, which is much too far north on the map, to Elmali, Böyük Tatlar, and Khurman Kalessi.

Khurman Kalessi, a proud castle possibly of early Turkish origin, stands on a crag just at the junction of the Maragos Tchai with the Khurman Su. Between Khurman Kalessi and Maragos there are three inscriptions in large letters on three huge rocks by the roadside ; taken together, they are the most interesting topographical documents within my knowledge. The first consists of eight heroic hexameters ; the second, of two hexameters ; and the third is an elegiac distich. Two of them cannot be reached without artificial help, which everywhere in Turkey it is difficult to obtain. Of these two we have photographs. The third *can* be reached, but only with danger to life or limb. The surface covered by the inscription is so great that only a few letters in each line can be read at a time ; this done, one must climb down and up again, it being impossible to move horizontally along the face of the rock. They read as follows : —

No. 57.

ΑΚΙΛΛΙΟΥΧΕΙΡΙСΟΦΟΥΑΛΕΞΑΝ
ΔΡΟΥΤΟΥΚΑΙΦΙΛΙΠΠΙΟΥ

ΤΗСΔΕΚΟΡΗСΚΟΠΙΗСΠΟΤΑΠΗΛΙΒΑΤΟΙΟΘΟΡΟΥСΑ[С]
ΑΘΑΝΑΤΩΝΒΟΥΛΗСΙΝΥΠΕΚΦΥΓΕΝΑΡΚΤΟΝΑΠΗΜΩΝ
ΔΙΧΘΑΔΙΗΙСΚΩΜΗΙСΙΦΙΛΙΠΠΤΙΟΥΑΡСΙΝΟΟΥΤΕ
ΟΥΤΟСΑΡΙΓΝΩΤΟСΠΡΕΙΩΝΟΡΟСΑСΤΥΦΕΛΙΚΤΟС
ΕΠΛΕΤΟΔʼΑΡСΙΝΟΩΙΜΕΝΕΔΕΘΛΙΑСΑΡΡΟΜΑΗΝΑ
ΤΩΙΔʼΑΡΕΠΙΠΡΟΧΟΗСΙΔΥΩΠΟΤΑΜΩΝСΟΒΑΓΗΝΑ
ΠΙСΤΟΙΔʼΑΛΛΗΛΟΙСΕΤΑΡΟΙΠΕΛΟΝΩΝΦΙΛΟΤΗΤΑ
ΑΡΡΗΚΤΗΝΠΑΓΟСΟΥΤΟСΑΠΑΓΓΕΛΛΟΙΚΑΙΕΠΕΙΤΑ

No. 58.

ΤΟΥΑΥΤΟΥΧΕΙΡΙΣΟΦΟΥ
ΕΝΝΕΑΤΟΙΠΕΤΡΗΘΕΝΕΠΙΚΡΗΝΗΝΣΟΒΑΓΗΝΩΝ
ΚΑΛΛΙΡΟΟΝΣΤΑΔΙΟΙΚΟΡΑΚΟΣΠΟΤΑΜΙΕΟΠΑΡΟΧΘΑΣ

No. 59.

ΤΟΥΑΥΤΟΥΧΕΙΡΙΣΟΦΟΥ
ΕΓΓΥΘΙΤΟΙΣΟΒΑΓΗΝΑΚΑΙΑΙΓΛΗΕΝΤΑΛΟΕΤΡΑ
ΗΝΔΟΛΙΓΟΝΣΠΕΥΣΗΙΣ/ΟΥΣ ᴧΙΕΚΚΑΜΑΤΟΥ

As I attach importance to the thorough understanding of these inscriptions, I add the text in small letters : —

No. 57.

'Ακιλλίου Χειρισόφου 'Αλεξάν-
δρου τοῦ καὶ Φιλιππίου.

τῆσδε κόρη σκοπιῆς ποτ' ἀπ' ἠλιβάτοιο θορούσα[ν]
ἀθανάτων βουλῇσιν ὑπέκφυγεν ἄρκτον ἀπήμων
διχθαδίῃς [ρ]ώμῃσι Φιλιππίου 'Αρσινόου τε.
οὗτος ἀρίγνωτος Πρείων ὄρος ἀστυφέλικτος.
ἔπλετο δ' 'Αρσινόῳ μὲν ἐδέθλια Σαρρομάηνα,
τῷ δ' ἄρ' ἐπὶ προχοῇσι δύω ποταμῶν Σοβάγηνα.
πιστοὶ δ' ἀλλήλοις ἔταροι πέλον, ὧν φιλότητα
ἀρρήκτην πάγος οὗτος ἀπαγγέλλοι καὶ ἔπειτα.

No. 58.

τοῦ αὐτοῦ Χειρισόφου.

ἐννέα τοι πέτρηθεν ἐπὶ κρήνην Σοβαγήνων
καλλίροον στάδιοι Κόρακος ποταμοῖο παρ' ὄχθας.

No. 59.

τοῦ αὐτοῦ Χειρισόφου.

ἔγγυθί τοι Σοβάγηνα καὶ αἰγλήεντα λοετρά·
ἢν δ' ὀλίγον σπεύσῃς [λ]ούσ[εα]ι ἐκ καμάτου.

[NOTE. The following changes in the text of these inscriptions have been adopted, chiefly in accordance with the suggestions of Professor F. D. Allen, since the paper was received from Dr. Sterrett: in No. 57, 1, κόρη σκοπιῆς for Κορησκοπίης, and θορούσα[ν] for θορούσα[ς]; 57, 3, [ῥ]ώμῃσι for κώμῃσι, and the period at the end of the verse; 57, 4, Πρείων ὄρος for Πρειώνορος. In No. 59, 2, [λ]ούσ[εα]ι for · · · · · ·.
They may be thus translated : —

No. 57.

Epigram of Acilius Chirisophus, the son of Alexander who is also called Philippius.

" Once on a time, by the counsels of the Immortals, a girl escaped unhurt from a bear which rushed down from this lofty crag, through the twofold strength of Philippius and Arsinous. This is the unshaken mount Prion, known to all men. The home of Arsinous was Sarromaëna ; that of Philippius was Sobagena, at the confluence of two rivers. They were faithful comrades, and may this rock declare their unbroken friendship even to future ages."

No. 58.

Of the same Chirisophus.

" It is nine stadia from this rock to the fair-flowing spring of Sobagena, on the bank of the river Korax."

No. 59.

Of the same Chirisophus.

"Near by is Sobagena with its bright clear baths. If you will hasten a little, you may bathe yourself after your toil."

In consequence of the changes in the text. Dr. Sterrett's interpretation of the first inscription has been omitted. — EDD.]

From this it is clear that Khurman Kalessi occupies the site of Sobagena, inasmuch as it is just at the junction of the Maragos Tchai with the Khurman Su. For a similar reason, we must conclude that the ancient name of the Khurman Su was Korax. We saw no spring at Khurman Kalessi ; the ancient spring may have dried up, or perhaps reference is made to the clear cold water of the Korax itself. Sarromaëna possibly was situated near Maragos, which name may even be a corruption of Sarromaëna. Nine stadia is about the true distance from Khurman Kalessi to the rock bearing the inscription.

Leaving these inscriptions, we passed by Maragos ; thence through an exceedingly mountainous country to the valley of Saris, near which

Coduzalaba must have stood. There are seven or eight villages in this plain, which I left unvisited, as this route had been taken by Messrs. Ramsay and Wilson a few years ago. Here we crossed the Antitaurus by a much less laborious pass than before. Our road lay by Ekrek, Karadaghi, where there is a good Seldjuk khan, and Zerezek (the old Arasaxa), to Caesarea. Hence back to Indjesu, and then through the wonderful volcanic region of Urgüp and Udjessar, the home of the ancient Troglodytes. We have numerous photographs from this region, which will no doubt be greeted as a valuable addition to science and archæology. Thence by Nev Sheher to the rock-cut dwellings of Tatlar. The cliffs of Tatlar have been so honeycombed by the excavating Troglodytes, that large boulders occasionally break off and thunder down like an avalanche on the unsuspecting village below, leaving death and destruction in their wake. Only four days before our visit such a fragment, weighing many tons, had precipitated itself upon the village, destroying twelve houses and killing five men. Several other boulders must soon fall. The women of the village had the idea that we had come to investigate the cliffs and take measures for the protection of their houses. As we roamed about, they would anxiously inquire of us : " Is there any danger for our house?" "Will that rock fall?" The enchanted book, of which Hamilton speaks, has been carried off by the Greeks of Arabsun, the corrupted form of the ancient Yarapason.

Hence our road lay north by Tuzkieui across the Halys to Hadji Bektash, the headquarters of the Dervishes of that name. The Halys is here a broad stream, easily forded in the summer season. North of Hadji Bektash lies an absolutely unknown district, which we explored as far as Pashakieui, to settle the courses of the rivers ; then from Pashakieui we followed the Delidje Irmak in its downward course to a point directly south of Böyük Nefezkieui, to which we now went.

Here I have to record one of the most important discoveries of the journey, but must preface it with a few historical remarks.

The ancient Tavium was a place rather insignificant in itself, although it is called the ἐμπόριον τῶν ταύτῃ; but the more important geographically, because it was the centre from which diverged seven roads, five of which are given in the Peutinger Table, and the remaining two in the Antonine Itinerary. Distances along these

roads were measured from Tavium; consequently it is of the highest
importance to discover its real site, for on it depends the geography of the whole country between Ancyra and Amasia. Tavium
has been placed by different scholars at Tchorum, Nefezkieui,
Boghazkieui, and Yozghad; but until recently those best entitled
to an opinion had settled on Nefezkieui as the true site. But
in November, 1883, Professor Hirschfeld of Königsberg published
an article "Tavium" in the *Sitzungsberichte der Akademie der
Wissenschaften zu Berlin*, in which he declined to accept for
Tavium any of the sites hitherto suggested. He tries to show by
arguments, which cannot be accepted, that the town must be sought
for on the left bank of the Halys, and that its site is occupied by
Iskelib, a town situated a whole degree north of Böyük Nefezkieui.
In January, 1884, Professor Kiepert published in the *Sitzungsberichte*
(as above) his *Gegenbemerkungen zu der Abhandlung des Hrn. G.
Hirschfeld über die Lage von Tavium*, from which it appears that he
is very loath to give up the site of Böyük Nefezkieui; but he finally
suggests Aladja, or a point immediately south-east of Aladja.

But now to my facts! In a cemetery between Böyük Nefezkieui
and Assara (on Kiepert's map, *Aksikara*), and immediately west of
the Acropolis of Böyük Nefezkieui, I found a Roman milliarium
which reads:—

No. 60.

IMP
NERVA CAESARAV////
PONT MAXTRIB POTESVII
COS III PP RESTITVIT
PERPOMPON//////M
BASSVM LEG PROPR

P MĪ A

Now this is the FIRST milestone from somewhere, MĪ being written
instead of the more common M̊;* but as distances were reckoned
here from Tavium, it is necessarily the first milestone on the road
from Tavium to Ancyra.

---

* Stones vary considerably in this respect, and we find M, MI, M̊, MIL, M̊P,
PM, M̊ILP.

Again, in the cemetery of Tamba Hassan, a village two hours north of Böyük Nefezkieui, on the road to Boghazkieui, there is another milliarium, badly defaced and almost totally illegible.

No. 61.

[This inscription has not been received. — EDD.]

Now, as I understand it, Tamba Hassan is none other than Tomba or Tonea of the Peutinger Table, the first station on the road from Tavium to Comana in Pontus. Hirschfeld points out that Tonea and Tomba are two names for the same place. The Table has Tonea XIII (and Tomba XVI) MP from Tavium, a distance which corresponds reasonably well with the site of Tamba Hassan.

I copied about twenty-five inscriptions at Böyük Nefezkieui, nearly all of which are Christian epitaphs of no historic value.

It has been stated that the ruins of Böyük Nefezkieui are too insignificant to represent Tavium. This is not the case. It is true that at the village itself there are only comparatively small fragments ; but all the villages around Böyük Nefezkieui are full of architectural fragments, and the cemetery, whence comes the milliarium, has scarcely any other stones in it except cippi, columns, and fragments of epistyles, of considerable size and weight. A future traveller will no doubt find the hot springs in the region between Böyük Nefezkieui and Yozghad. At Kütchük Nefezkieui there is a large spring of cold water, formerly used for baths ; part of the bath-house still exists. I found only Roman coins at Nefezkieui, of the Caesarean coinage. The soil is very fertile, and yields abundant harvests of wheat ; and the people plant nothing else. From all this it would seem that there can no longer be a reasonable doubt concerning the site of Tavium.

From Nefezkieui we visited Boghazkieui and Üyük, taking photographs of the celebrated rock sculptures. Leaving Üyük, we passed by Sungurlu, Beshbunar, Aghabunar, and Barshili, to a point on the Halys one hour south of Kaledjik, and crossed the mountains to Ancyra. The results of this part of our journey were purely chorographical. In the cemetery of a village two hours east of Ancyra, on the Enguri Su, I copied the one hundredth milestone from Tavium (No. 62). I was much hurried at the time, and failed to get the name of the village ; but I shall learn it hereafter through a friend in Ancyra.•

No. 62.

```
IMP CAESVAI
SEVERO
ROPIOFEL·IN
VICTO AVG·TRIB
POT II COSI
```

M P

At Ancyra we were compelled to consider our journey finished, scientifically speaking. It was necessary for Mr. Haynes to reach Nicomedia by a certain day, and our one thought henceforth was to travel westward as rapidly as possible. The inscriptions copied on the journey number three hundred and fifty. Mr. Haynes took three hundred and twenty photographs. The route-surveys are in the hands of Professor Kiepert.

J. R. S. STERRETT.

ATHENS. October, 1884.

www.ingramcontent.com/pod-product-compliance
Lightning Source LLC
Chambersburg PA
CBHW022028190326
41519CB00010B/1633